Citrus Engineering Conference 2014

Lake Alfred, Florida, USA
12 June 2014

ISBN: 978-1-5108-2501-7

Printed from e-media with permission by:

Curran Associates, Inc.
57 Morehouse Lane
Red Hook, NY 12571

CURRAN ASSOCIATES INC.
proceedings.com

Some format issues inherent in the e-media version may also appear in this print version.

Printed by Curran Associates, Inc. (2016)

Citrus Engineering Conference
c/o Jose Reyes
700 Experimental Station Road
Lake Alfred FL 33850

Phone: (863) 956-1151
Fax: (863) 956-4631

jireyes@ufl.edu

Additional copies of this publication are available from:

Curran Associates, Inc.
57 Morehouse Lane
Red Hook, NY 12571 USA
Phone: 845-758-0400
Fax: 845-758-2633
Email: curran@proceedings.com
Web: www.proceedings.com

TRANSACTIONS
of the 2014

SETTING THE STANDARD

CITRUS ENGINEERING CONFERENCE

Sponsored by

FLORIDA SECTION

AMERICAN SOCIETY of MECHANICAL ENGINEERS

LAKE ALFRED, FLORIDA
JUNE 12, 2014
VOLUME LVIII

Global Communities
Knowledge and Community Sector

June 12, 2014

Dear Colleagues,

The Citrus Engineering Conference was initially organized in 1954 by Andy Hines of Florida Power (now Duke Power), to promote an exchange of information about the Citrus industry. The tenets of these conferences have kept pace with the continuing needs of the industry. The goals of the Conference are kept alive by the citrus industries staunch industrial support for the conference allowing the Citrus Engineering Conference Committee to arrange for the engineers, researchers, and managers who are dedicated to strengthen the industry to meet today to promote the continuing improvement of citrus products.

The Conference will further the goals of ASME by promoting the art and science of Mechanical Engineering. It will also provide a great opportunity for professionals in the citrus industry worldwide to meet, network, and exchange ideas on management and technology. The conference is supported within the ASME by the ASME's Process Industries Division chaired by Florin Iancu, the ASME Florida Section, chaired by George Cera and the ASME Florida Section Citrus Engineering Technical Chapter chaired by Matthew Smolowitz.

While the conference has changed in venue and format over the years, the engineering and research topics presented at the conference have made this the best Regional industry conference in ASME. The timeliness of the information presented today will keep the Florida Citrus Industry in the forefront of improved product quality and productivity. ASME is proud to sponsor this distinguished event.

Chair Dr. Elizabeth Webb and the Members of the Citrus Engineering Conference Committee have worked hard and diligently to organize the topics and provide the facilities to make this conference valuable to all who attend. They are to be commended for their efforts.

Welcome to the 2014 Florida Section ASME Citrus Engineering Conference on behalf of the ASME.

Richard H. Bunce, P.E.
Vice President Global Communities

The American Society of Mechanical Engineers

2014 CITRUS ENGINEERING CONFERENCE PROGRAM

8:00 a.m.	**REGISTRATION:** Lobby, Ben Hill Griffin Hall, CREC
8:30 a.m.	**WELCOME:** Elizabeth Webb, Brown International, Committee Chair and Co-Moderator Ian Houston, JBT FoodTech, Conference Co-Moderator
8:40 a.m.	**EXPANDING THE HORIZON OF CITRUS FIBER** Brock Lundberg, Ph. D., Fiberstar Bio-Ingredient Technologies Inc, Eau Claire, Wisconsin
9:20 a.m.	**ENGINEERED AIR CURTAINS AND AIR CURTAIN VESTIBULES FOR HIGH TRAFFIC COOLER AND FREEZER DOORWAYS** Pete Smith, HCR Incorporated, Lewistown, Montana
10:00 a.m.	**REFRESHMENT BREAK**
10:25 a.m.	**TROPICANA'S OPTIMIZATION OF LANDFILL METHANE GAS USE** Glenn Johnson, Tropicana Manufacturing Co. Inc, Bradenton, Florida
11:05 a.m.	**TO TREAT OR NOT TO TREAT CITRUS WASTEWATER: WHAT ARE THE ALTERNATIVES?** Denise Johnston, Veolia Water Solutions and Technologies, Pennsauken, New Jersey
11:45 a.m.	**LUNCH - Catered barbeque**
1:10 p.m.	**INTRODUCTION OF ASME OFFICIALS**
1:20 p.m.	**31ST CITRUS ENGINEERING AWARD** Presented by Dayton Amey, Tropicana Products
	• DOOR PRIZE DRAWING •
1:35 p.m.	**GREENING PANEL DISCUSSION - OUR FUTURE WITH HLB: WHERE TO FROM HERE?** Moderator: Dan King, Director of Scientific Research, Florida Department of Citrus Panelists: Elizabeth A. Baldwin, Supervisory Research Horticulturalist, USDA-ARS; Mike Irey, Director of Research, US Sugar/Southern Gardens Citrus; Dana D. Stump, President, Direct Driven Solutions LLC; Jim Graham, Professor of Soil Microbiology, UF/IFAS CREC; Greg McCollum, Research Physiologist, USDA-ARS
2:45 p.m.	**REFRESHMENT BREAK**
3:00 p.m.	**GREENING PANEL DISCUSSION, CONTINUED**
	• DOOR PRIZE DRAWING •

2014 CONFERENCE COMMITTEE

ELIZABETH WEBB, PE – Committee Chair
Technical Service Manager
Brown International Corp. LLC. Winter Haven, FL

MATTHEW SMOLOWITZ – Chair, ASME Florida Section Citrus Engineering Technical Chapter
Senior Engineer, Gas Turbine Service Engineering
Siemens Energy, Inc., Orlando, FL

PETE VAN CAULIL – Treasurer
Manager, Corporate Commodity Engineering
Siemens Energy, Inc., Orlando, FL

BRETT STEGMAIER – Secretary
Engineering Manager
Florida's Natural Growers, Inc., Lake Wales, FL

CLARISSA ALBARRAN
Senior Sales Engineering Manager
Brown International Corp. LLC. Winter Haven, FL

RICHARD BUNCE, PE
Principal Engineer, Materials and Technology
Gas Turbine Engineering
Siemens Energy Inc., Orlando, FL

MARJORIE DRAKE
Senior Engineer, Environmental Operations
Site Resource, HSE
Tropicana Manufacturing Co. Inc., Bradenton, FL

DEAN ELLERBROCK
Corporate Refrigeration Manager
Wayne Farms LLC, Jack Alabama

IAN HOUSTON
Process Engineering Manager, Juice Processing Technologies
JBT FoodTech, Lakeland, FL

MUHAMMAD M. RAHMAN
Professor and Director of Graduate Programs
Department of Mechanical Engineering
University of South Florida, Tampa, FL

JAMES STANA
Past Chair, ASME Florida Section
JMS3DPrinting.com, Orlando, FL

2014 CITRUS ENGINEERING AWARD

RALPH L. HAYES, PhD

Ralph Hayes was born in Steubenville, Ohio and served in the Army during the Vietnam era. After serving in the Army, Ralph and his wife moved to Columbus, Ohio. The first 10 years of his career were spent furthering his education and working in the pulp and paper industry as a design and systems project engineer. Ralph has earned his BS in Mechanical Engineering, a MBA, and a PhD in Business Management

Ralph moved to Florida in 1978 and went to work in the citrus industry as an engineering manager for Tropicana Products, Inc. in Bradenton, Florida. There he progressed through various levels of increasing responsibility. In 1988 he continued to advanced his career by holding various engineering and senior management level positions within the citrus industry, including: Engineering Manager, Tropicana Products, Inc.; Engineering Manager, The FMC Corporation; Director of Engineering, SeparaSystems LP (a joint venture between FMC and DuPont); Senior Account Manager, The FMC Corporation; Vice President of Operations and Engineering, Golden Gem Growers; Division Vice President, A. Duda & Sons (Citrus Belle).

Ralph is currently working in the following positions: President and CEO, Natural Citrus Products Corporation; President and Owner, Hayes Enterprises; Adjunct Professor; Keller Graduate School of Management, DeVry University. He is has also served in the following professional associations: Citrus Engineering Conference Committee; Board of Directors,

Florida Citrus Processors Association; National Juice Products Association; Association For Facilities Engineering (CPE).

Ralph has traveled and worked extensively in the US and abroad, providing engineering, technical services, project management and operations management services and training to the pulp and paper industry, and for the past thirty plus years service to the citrus industry.

When reflecting on his career, Ralph says that there are few aspects of this great industry that he have not dealt with, and in all that he does in dealing with customers, as well as with employees, he has always strived to achieve a proper balance between his technical, management and leadership skills.

2014 CITRUS ENGINEERING FOUNDER AWARD

ANDY HINES

The 2014 Citrus Engineering Conference would like to recognize Andrew Hines, Jr. for taking the initiative 60 years ago to commence this forum for starting and keeping the Citrus Industry engaged in actively sharing information. As a result of his personal efforts, the Citrus Engineering Conference has grown, developed, and endured for decades. We look forward to continuing the tradition and benefitting directly from Andy's work for many more years to come.

PREVIOUS CITRUS ENGINEERING AWARD RECIPIENTS

Hugh W. Schwarz	1982
Thomas B. Kimball	1983
William Belk	1984
Ralph Cook	1985
Don H. Kimball	1986
Charles S. Walker	1987
John W. Seabury	1988
Llewellyn D. Fain, Jr.	1989
Cedric D. Atkins	1990
Maurice L. Warner	1991
John M. Cahill	1992
Robert P. Robbins	1993
Robert D. Keller	1994
Harold Heath	1995
Larry Kelly	1996
James W. Kesterson	1997
Harold Rebeck	1998
D.L. "Don" Wiggins	1999
Guillermo T. Segredo	2000
Chin Shu Chen	2001
Fred T. Fulks	2002
Donald W. Wise, Jr.	2003
Carlos Odio	2004
Jodie D. Whitney	2005
Ron Bushman	2006
Michael Suter	2007
William M. Miller	2008
Robert J. Braddock	2009
Roger D. Waters	2010
Robert B. Johnston	2012

CONTENTS

EXPANDING THE HORIZON OF CITRUS FIBER

Brock Lundberg, Ph.D.
President/CEO, Fiberstar Bio-Ingredient Technologies, Inc.
2322 Alpine Road, Suite 2
Eau Claire, WI 54703
Email: lundberg@fiberstarbio.com

While fibrous byproducts, such as citrus fibers, are abundant, using them in food products to improve food nutrition and quality while not degrading taste or texture can be challenging. To incorporate citrus fibers into foods, they need to be processed in such a way that they impart minimal degradation of taste and texture in food products. In designing the product and production process, the final applications for the products need to be evaluated early in the product development cycle. Additionally, to better understand their functionality in food products, their properties, composition, and rheology need to be better understood. A specific aim of this study was to characterize the properties, composition, and rheology of a particular range of citrus fibers produced by Fiberstar, Inc.

Applications for citrus fiber products are broad and vary from bakery and meat to dairy and beverages. In beverage products, citrus fibers function to improve stability and/or texture. In meats, citrus fibers bind both water and oil to increase yields and can replace synthetic salts and stabilizers. Other applications include partially replacing oil, fat, and eggs to reduce costs in bakery, meat, and dairy products. The typical usage rate is less than 1% but citrus fibers can bind ten times its weight which frequently equates to both a cost savings and texture benefit.

The flow properties of citrus fiber solutions are clearly non Newtonian and the type of model that best fit the citrus fiber varied depending on its particle size. Particle size of citrus fiber solutions also significantly changed their apparent viscosity. As the citrus fiber hydrates, the fiber structure swells as it absorbs water into its cell structure.

ENGINEERED AIR CURTAINS AND AIR CURTAIN VESTIBULES FOR HIGH TRAFFIC COOLER AND FREEZER DOORWAYS

Pete Smith, President
HCR Inc.
80207 Highway 87W
Lewistown, MT 59457
Email: Pete@hcr-inc.com

It is a known fact that often 50% or greater of the total refrigerated load at a refrigerated warehouse is attributed to doorway infiltration. Since the single largest contributor to the power bill in a cold storage facility is the refrigeration system, reducing doorway infiltration by even a small amount directly contributes to the bottom line. The key to reducing this expensive load is to reduce door open time and or reduce the infiltration while the door is open by some type of “air door” device. This presentation will look at doorway devices such as stand alone traffic doors, traffic doors with air doors (hybrids) and stand alone air doors including conditioned air curtain vestibules.

Understanding in detail how a doorway is used is critical and is the most difficult part of doorway application engineering and is the challenge to all in the industry. In addition, the end user has to sort through claims and promises made by all manufactures. This presentation looks at third party performance validation which is available to all traffic door and air door manufactures. The final segment of this presentation will look at all available doorway options to the cold food storage and processing industry.

The paper that follows includes a more in-depth analysis from a third party performance validation, and includes the method of how doorways were modeled and the engineering behind the benefits stated in this presentation.

Evaluation of Horizontal Recirculatory Air Curtain Efficiencies - Cooler to Conditioned Space.

Daniel J. Rhyner, BSME
HCR, Inc.

Abstract

Air curtains have long been used to reduce losses from high traffic doorways in cold storage facilities. Different varieties of air curtains have been used including vertical (down blast) non-recirculatory, horizontal non-recirculatory and horizontal recirculatory. The effectiveness of air curtains to provide thermal separation between rooms of differing temperatures has long been a matter of debate. Previous experiments have been done to determine the effectiveness using tracer gas decay methods. This paper will evaluate the thermal effectiveness of the horizontal recirculatory air curtain when applied in a typical cooler application using an environmental chamber. This allowed testing of the air curtain's ability to reduce infiltration on the actual thermal envelope. A vertical non-recirculatory air curtain was also evaluated for comparison purposes. The air curtains were tested in an environmental chamber that was divided into two rooms of differing temperature. The chamber was instrumented to measure all energy transfer into and out of each room. The cold room was maintained at a temperature of 4°C (39°F) and the warm room was maintained at a temperature of 24°C (75°F) and 60% RH throughout all tests. Three phases of testing were conducted for each air curtain to completely evaluate all energy losses – a completely closed doorway test for calibration, a completely open doorway test with no air curtains and an open doorway test with the air curtains adjusted and running. Completion of the testing revealed the horizontal recirculatory air curtain had a thermal efficiency of 71% while the vertical non-recirculatory air curtain had an efficiency of 38%. The results were slightly lower than theoretical calculations which may be attributable to chamber size. Additional tests were conducted using a smaller doorway for comparison. Applying the results from the smaller doorway yields an effectiveness of 76.2%. Using the theoretical open doorway energy loss yields an effectiveness of 80%

Introduction

Air curtains are generally used to reduce infiltration to both cold and warm spaces. They are used extensively in the cold storage industry to provide thermal separation and minimize refrigeration loads for coolers of differing temperature, freezers of differing temperature, freezers to coolers, coolers to ambient areas, and freezers to ambient areas. There have been several tests done in recent decades to determine the effectiveness or performance of different air curtain configurations at differing temperatures; however, most if not all of these tests were conducted using a tracer gas decay method. A summary of results from a few of these tests is summarized in Table 1. (Downing and Meffert 1993, Pham and Oliver 1983, Longdill and Wyborn 1979, Valkeapaa et al. 2005). For single air curtains in static cooler applications the results vary from as low as 42% to as high as 82%.

The current method for determining the effectiveness of doorway protective devices such as air curtains as recommended in the 1990 ASHRAE Handbook-Refrigeration is to take the ratio of the infiltration at the doorway with the protective device in place and the infiltration of the doorway with no protective device in place calculated from the Gosney and Olama model (ASHRAE 2002) for air infiltration and subtract from 1. The 2002 ASHRAE Handbook-Refrigeration states the maximum effectiveness of air curtains to be 0.7.

Table 1. Previously reported effectiveness values – single air curtains in cooler applications.

Air Curtain Type	Effectiveness	Test Method	Source
Vertical non-recirculating	79% ±3 42% ±15 68-83%	Tracer Gas Decay	Pham and Oliver (1983) Downing and Meffert (199[illegible] Longdill and Wyborn (1979
Horizontal non-recirculating	42%	Tracer Gas Decay	Valkeapaa et al. (2005)
Horizontal recirculating	76 ±3 82%	Tracer Gas Decay	Pham and Oliver (1983) Longdill and Wyborn (197[illegible]

No previous data was found where air curtains were tested by trying to measure actual heat and moisture gains or losses at the doorway. This is understandable considering the difficulties of measuring infiltration velocities in three dimensions at an existing facility not to mention the temperature and humidity measurement grid required. A solution to this was to do a calorimeter test in an environmental test chamber large enough to be representative of typical industry applications. A calorimeter test allows the evaluation of all energy losses at the doorway with and without a doorway protective device in place. The effectiveness is then determined by the ratio of the measured energy loss at the doorway with the protective device in place to the measured energy loss at the doorway with no protective device in place and subtract from 1.

The cold storage industry today still does not recognize a particular number for air curtain efficiency. There is still a need for better information regarding these efficiencies especially now with a focus on sustainability and green technologies. The main objective is to begin providing those numbers starting with single air curtains in a static cooler application with a large enough opening and a significant temperature differential to provide meaningful results. A static condition for the purpose of these tests is defined as infiltration at a doorway due to temperature differences only. No pressure imbalance exists to create additional infiltration.

Experimental Method

According to ASHRAE the effectiveness of a protective device is defined by the following equation: $E = (1 - I_p/I_u) \times 100\%$ (1)

where

I_p = protected doorway air exchange rate (ft^3/hr)
I_u = unprotected doorway air exchange rate (ft^3/hr)

I_u is calculated from the Gosney and Olama model (ASHRAE 2002) to be

$$I_u = 795.6\ (A)\ (1-\rho_i/\rho_r)^{0.5}\ (gH)^{0.5}\ (Fm) \quad \text{in ft}^3/\text{hr} \qquad (2)$$

Where

795.6	=	3600 s/h x 0.221;
A	=	doorway area, ft^2;
ρ_i	=	density of infiltration air, lbm/ft^3;
ρ_r	=	density of refrigerated air, lbm/ft^3;
g	=	gravitational constant, 32.17 ft/s^2;
H	=	doorway height, ft; and
Fm	=	density factor.

The density factor is defined by Gosney and Olama (ASHRAE 2002) as

$$Fm = [2/(1+(\rho_i/\rho_r)^{1/3})]^{3/2} \quad (3)$$

This method of calculating the doorway protective device effectiveness is fine for a fixed barrier since the effectiveness is easily calculated using standard heat transfer equations and applying a leakage factor; however, it does not work as well for air curtains because they are more complex systems to evaluate. In proper determination of effectiveness of air curtains, all refrigeration losses should be accounted for and thus should include all sensible and latent heat gains or losses from the infiltration. The theoretical effectiveness calculation of the air curtain becomes a direct calculation of the ratio of the calculated doorway refrigeration loss with the air curtain in place and the calculated open doorway refrigeration loss.

$$E = (q_o\text{-}q_a/q_o) \text{ x } 100\% \quad \text{or} \quad (1\text{-}q_a/q_o) \text{ x } 100\%$$

q_a = energy loss with air curtain, Btu/hr or kW
q_o = energy loss of open doorway, Btu/hr or kW

The open doorway energy loss is determined by using the Gosney and Olama model for fully established flow (ASHRAE 2002) which is the same as equation (2) above but multiplied by density of the refrigerated air and the enthalpy difference of the infiltrating air and the refrigerated air. Thus the equation for refrigeration loss of an open doorway becomes

$$q_o = 795.6\ (A)\ (h_i - h_r)\ \rho_r\ (1\text{-}\rho_i/\rho_r)^{0.5}\ (gH)^{0.5}\ (Fm)\ (D_f) \quad (4)$$

The theoretical refrigeration loss for the horizontal recirculatory air curtain is more complicated and cannot be shown here but the calculations and results used for theoretical effectiveness calculations is included in Appendix A. The theoretical effectiveness of a horizontal recirculatory air curtain placed on a 7-ft by 7-ft opening between a 39°F cooler at 90% RH and a 75°F area at 60% RH is calculated to be

q_a = 33,840 Btu/hr
q_o = 174,000 Btu/hr

$$E = (1 - 33{,}840/174{,}000) \text{ x } 100\% = 80.6\%$$

q_o is calculated using a doorway flow factor of 0.80 per Hendrix et al. (1989) (ASHRAE 2002).

The requirement was to determine the experimentally measured effectiveness of air curtains instead of the theoretical effectiveness. The same effectiveness equation applies but q_a and q_o become measured values using the calorimeter environmental chamber.

Test Equipment

A calorimeter capable of producing a wide range of environmental conditions was used to evaluate the performance of the horizontal recirculatory air curtain. An insulated wall with a 7-ft by 7-ft opening was centered between the air handling equipment on both sides of the chamber.

Figure 1 shows a schematic of the environmental chamber and the components contained inside.

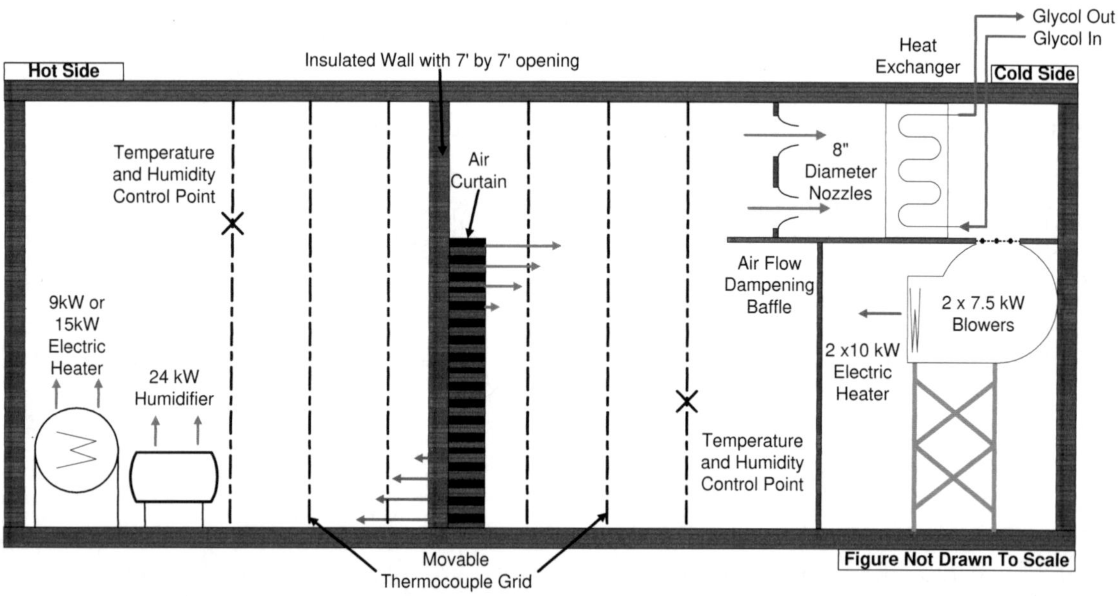

Figure 1. Environmental chamber schematic.

Cooling was supplied to the cold side heat exchanger by circulating ethylene glycol from two air cooled chillers each rated at 17 tons of cooling capacity. For the majority of tests, only one chiller was used and the cooling load on the chamber was varied as necessary through modification of flow rates, reduction in number of chiller compressors in use, and the addition of heat to the chiller loop. Airflow dampening baffles were placed in front of the blowers on the cold side to prevent air flow in the direction of the air flow though the opening. The cooling load on the chamber can be calculated based on the measurement of refrigerant loop mass flow rate, the temperature change between the cold side heat exchanger inlet and outlet, and the specific heat of glycol using the formula $Q=mc_p\Delta T$. It was necessary to derive the glycol concentration from the solution density and use this value along with the average solution temperature for a given test to derive its specific heat using interpolations of values from ASHRAE Standard 21.6. The power input to the cold side was also measured using watt transducers. This includes power going to the blowers, fans and VFDs. The power going to the air curtain was measured by a separate watt transducer. The chamber was calibrated and all heat coming through the walls was also measured using thermocouples on all inner and outer wall, floor, and ceiling surfaces.

Heating was provided to the warm side by a 9kW or a 15kW heater connected to a variable frequency drive which was controlled to maintain the chamber temperature using the Agilent VEE Pro data acquisition program. Humidification was achieved through the use of a steam generator which was also varied by a variable frequency drive and controlled using the data logger. Several box fans were placed near the air handling and temperature and humidity control equipment and were arranged to blow parallel to the opening to encourage air circulation on the cold and hot side of the chamber while not interfering with energy and air flow from the warm side to the cold side. The fans were positioned to help ensure a uniform thermal reservoir on each side. All heating loads on the warm side were measured using watt transducers. These measured loads included the heaters, steam generator, VFDs, fans, lights, etc. All electrical heating inputs were accounted for.

Airflow measurements were taken at the 7-ft by 7-ft opening with all air handling equipment and fans running but with no temperature difference between the sides to ensure that the air flow created by these devices had a negligible effect on the air flow from side to side. All air flow measurements were well below the 40 ft/min limit commonly found in testing standards.

The acquisition of planar air temperature profile data across a 17-ft by 12-ft cross section perpendicular to the opening was accomplished using a type T thermocouple grid configured so as to move along the length and width of the chamber while being able to maintain planar form. The grid was composed of 6 vertical columns each having 8 thermocouples attached at 1.5-ft vertical intervals from top to bottom for a total of 48 thermocouples. A spacing of 3-ft was maintained between each vertical column for every test. Three temperature plane locations were used for each test to give a 3D representation of the temperature profile.

A data logging program was developed in Agilent VEE Pro to monitor test conditions in real time. All sensors types shown in Table 2 were connected to three HP data loggers controlled by this program. The output of the program provides sufficient information to formulate a comprehensive energy balance on the chamber including all heating and cooling loads. It also includes the planar temperature distribution, cold and hot side dew point readings, and differential pressure readings across the cold side heat exchanger and air flow measuring nozzles. All data was exported to Microsoft Excel for further analysis and visualization. Contour plots were developed for planar temperature profile visualization with data playback capability to enable the user to quickly visualize how any planar temperature profile data set changes over time. A data sampling rate of approximately one data set per nine seconds was achieved with the primary data logging program. A secondary program was developed for monitoring only the planar temperature profile which had a data sampling rate of approximately one data set per three seconds. The faster sampling program was used to develop time lapse videos of the temperature distribution when the air curtain was turned on and off.

Table 2. Instrumentation list

Measurement	Instrument	Brand	Range	Accuracy	Description/Location
Temperature	Type T Thermocouple Probe	Omega	-250 to 220°C	1°C or 0.75% above 0°C 1.5% below 0°C	Glycol temperature measurements
Temperature	Type T Special Limits of Error Welded Thermocouple	Omega	-250 to 350°C	Greater of 0.5°C or 0.4%	Air temperature measurements
Dew Point	Model D-2-SR 2 stage chilled mirror sensor	General Eastern	-25°C to 85°C (up to 100% RH)	±0.2°C	Hot side (E6), Cold Side (South Blower), Cold Side Nozzle
Air-side Pressure Drop	Differential Pressure Transducer Model 26512R5WD1 1T1F	Setra	0 to 2.5" WC	±0.25% F.S.	Pressure drop across cold side heat exchanger
Glycol Flow Rate/Density	Mass Flow Sensor Model CMF200	Micro Motion	0 to 24.2kg/s	+/-0.05% of flow rate/ +/-0.0002 g/cc	Glycol (liquid) refrigerant loop mass flow rate/density
Watt Transducer	Model GW5-019D	Ohio Semitronics	0-2kW	±0.2% F.S.	Lights, fans, and other 120V devices
Watt Transducer	Model GW5-006E	Ohio Semitronics	0-120kW	±0.2% F.S.	Heaters, Blowers, Humidifiers on hot and cold sides
Watt Transducer	Model GW5-024E	Ohio Semitronics	0-16kW=4-20mA	±0.2% F.S.	Air Curtain

Testing Conditions and Procedure

Three phases of testing were performed to ascertain the performance of the air curtains. Table 3 gives a basic description of the chamber setup for each test. A full chamber open door calibration test was run to determine the heat loss through the chamber walls, ceiling, and floor not including the central wall. This test consisted of using only box fans and blowers to heat the entire chamber to steady state conditions. Using inner and outer wall, floor, and ceiling temperature sensor readings combined with watt transducer readings for the total heat input into the chamber and the dimensions for the chamber, a relationship was developed to calculate the heat loss or gain through the walls of the hot and cold sides of the chamber based on measured temperature differences in future tests.

A closed door calibration test was also completed with both warm and cold rooms brought to steady state conditions. There was no measurable loss from the warm room due to similar temperatures inside the warm room and the surrounding building. The measured loss for the cold room was 2.2kW. This test allowed for the calculation of heat transfer through the central wall in the chamber through proportional comparison of heat loss, temperature differential, and its associated surface area relative to the full chamber test.

The test conditions at which the open door, horizontal flow air curtain, and down blast air curtain were tested consisted of a cold side temperature of 39ºF and warm side temperature of 75ºF with 60% relative humidity. The warm side and cold side temperature and humidity control points were located on the outer edge of the measurement region relative to the opening with each being 8.2-ft from the opening. The cold room temperature control point was located 4.6-ft off the floor of the cold room while the warm room temperature/humidity control point was located 4.6-ft from the ceiling of the warm room. The testing conditions and iterations are shown in Table 3.

Table 3. Test conditions

Doorway Status	Doorway Size	Cold Side Set Point	Warm Side Set Point	Air Curtain Status
*Open	7-ft x 7-ft	Steady State	Steady State	Off
*Closed	7-ft x 7-ft	Steady State	Steady State	Off
Open	7-ft x 7-ft	4ºC	24ºC	Off
Open	7-ft x 7-ft	4ºC	24ºC	HCR On
Open	7-ft x 7-ft	4ºC	24ºC	Mars On

*Chamber Calibration Tests

The horizontal recirculatory air curtain was factory adjusted for optimal air flow velocity and air deflection angle and was tested at the standard operating condition. The unit was installed on the cold side to simulate a typical installation found in warehouse stores where the customer walks into a cold room to access perishable products.

A final test was completed to compare the horizontal, recirculatory air curtain performance to that of a down blast air curtain with no recirculation. A Mars high velocity industrial air curtain (Model HV84-2UH-BG) was tested. The down blast air curtain was installed by carefully following the manufacturer's instructions in the same way an end user would install the unit. The air curtain was mounted on the cold side of the chamber with the nozzle outlet raised one inch above the top of the doorway. The nozzle was configured using an angle template to blow air at a 15° angle outward from a vertical position so as to blow air slightly away from the doorway. The model provided was designed to have the greatest performance for doorways of a height between 10 and 12 feet. Because the doorway being tested had a height of 7 feet, it was necessary to significantly reduce the power supplied to the air curtain blowers. A VFD was installed and adjusted along with the air intake louvers to attain a condition where the air flow just reached the floor as specified in the operator's manual (Note: there is no VFD used in the operator's manual). A VFD setting of 30Hz was found adequate to meet the flow recommendations found in the installation manual. Tests were later run with the air curtain at VFD settings of 27.5Hz, and 25Hz to see the effect of air flow rate reduction although the end user would have limited means to accomplish this and determine the benefit in the field.

Data and Results

The effectiveness of the air curtain is defined by the difference of the heat transferred through the open door and the heat transferred through the air curtain divided by the heat transferred through the open door. The total heat input measured on the warm side minus the heat lost through the walls equals the heat transferred through the opening or air curtain if running. The cold side balance is used as a check for this hot side heat transfer calculation. The completely unprotected 7-ft x 7-ft opening was first tested at the standard operating conditions and the total heat transferred through the opening was found to be 35.7kW. The temperature profile shows the considerable flow of air and heat between the sides with the horizontal temperature gradients that are visible in Figure 2.

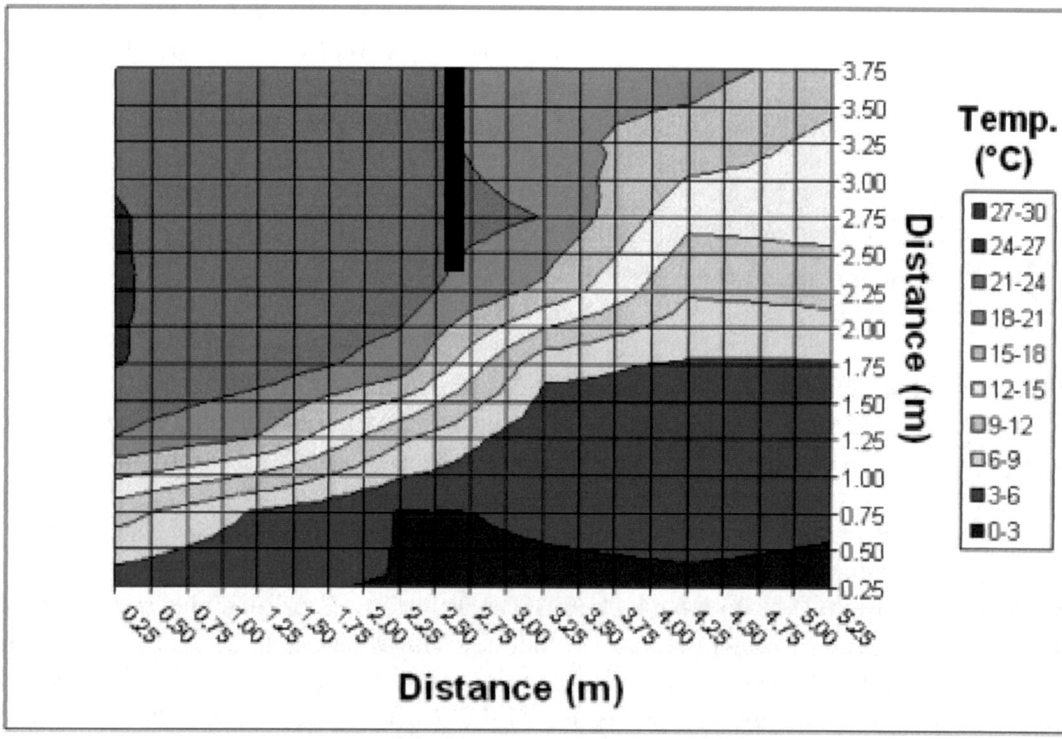

Figure 2. Temperature profile for open door test.

The flow of the warm air from the warm side through the top half of the opening and the flow of the cold air back from the cold side through the lower half of the opening are clearly visible in the 3D planar data taken during the open door test shown in Figure 3.

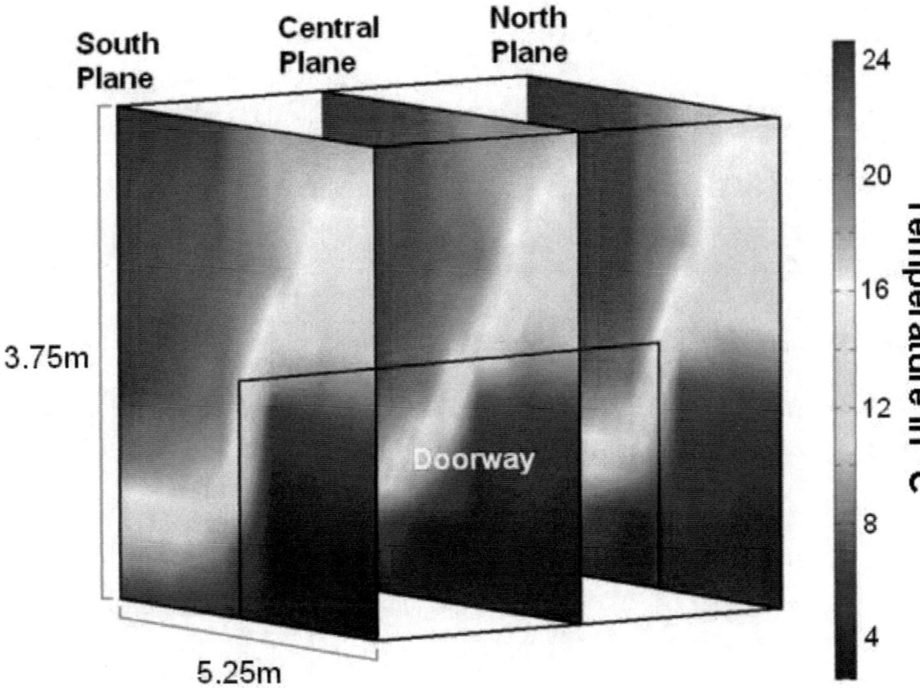

Figure 3. Planar temperature profiles with open doorway.

The horizontal recirculatory air curtain was adjusted for optimal air flow velocity and air deflection angle and was tested at the standard operating condition. With the air curtain running, the heat transferred through the opening was found to be 10.3kW. When compared to the open doorway, an effectiveness of 71% was obtained for the re-circulatory air curtain. Figure 4 and Figure 5 show this significant heat transfer reduction with the vertical temperature gradients. It is clear that the warm air is largely kept on the warm side and that the cold air is largely kept on the cold side.

Figure 4. Temperature profile for horizontal flow, recirculatory air curtain.

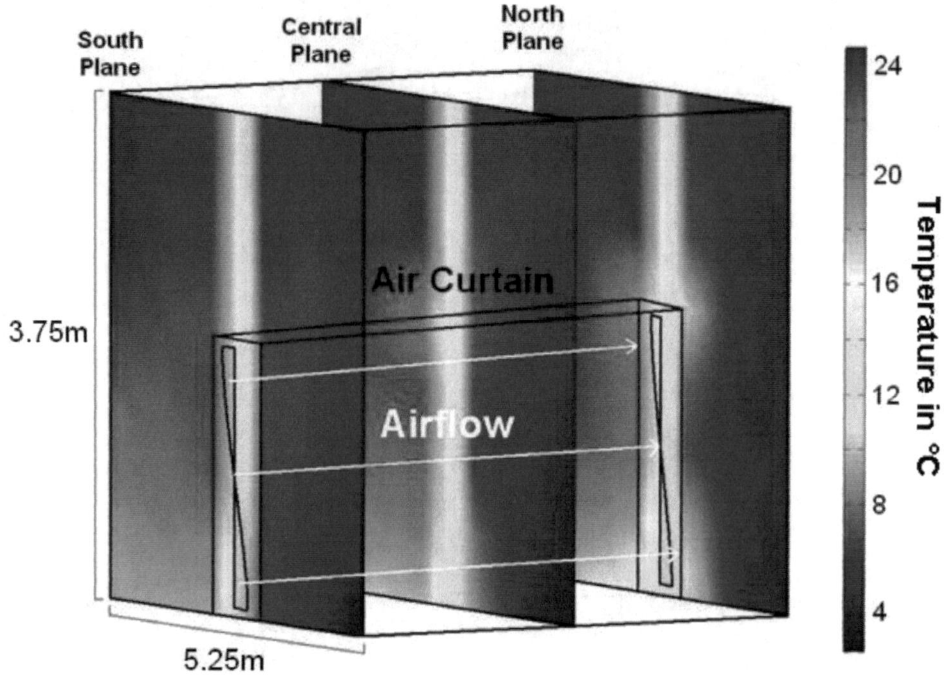

Figure 5. Planar temperature profiles for horizontal recirculatory air curtain.

It is worth noting that the measured open doorway heat loss was approximately 70% of the calculated theoretical open doorway heat loss using a doorway flow factor, Df of 0.8. Downing and Meffert (1993) as well as Pham and Oliver (1983) noted that a low ceiling height and obstructions near the door could affect measured open doorway infiltration air. The calorimeter rooms were significantly smaller than typical cold storage facilities and the ceiling heights much lower. The room height and size seem to have a great effect on heat loss due to recirculation of air back through the doorway. Upon completion of the air curtain testing, a smaller 3-ft by 3-ft opening was used between the warm and cold rooms to see if the size of the opening compared to room size or ceiling height had an impact on open doorway results. The calculated heat loss for the 3-ft by 3-ft opening yields 5.8kW. The measured value of open doorway heat loss was 4.9kW which is 85% of the calculated value. This is closer to theoretical than the 70% of calculated for the 7-ft by 7-ft opening. If the air curtain effectiveness is recalculated using 85% of the calculated value for the 7-ft by 7-ft opening, the air curtain effectiveness result is 76.2%. If we are applying air curtains to very large rooms with high ceilings we could expect the theoretical model for open doorway infiltration to match measured results based on prior testing and empirical data. With this being the case the final horizontal recirculatory air curtain effectiveness would be 80%.

The single vertical non-recirculatory air curtain was tested under the standard operating conditions with a VFD setting of 30Hz. The total heat transferred with the air curtain running was 22.1kW yielding a down blast air curtain effectiveness of 38% using the open doorway value of 35.7kW previously noted. Further optimization is possible although unlikely in the field.

Condensation was observed on the down blast air curtain housing and on the floor at the base of the opening directly below the air curtain. This was not observed at all with the horizontal recirculatory air curtain.

Discussion and Conclusion

The final results from the calorimeter testing indicate that the measured effectiveness of a horizontal recirculatory air curtain in a static cooler application with a temperature differential of 36°F is 71%. The vertical non-recirculatory air curtain had a much lower measured effectiveness applied in the same conditions at only 38%. There is clear visual evidence seen in the temperature profiles of the ability of a single horizontal recirculatory air curtain to provide an effective barrier between rooms of differing temperatures. In addition to the energy savings there is a significant benefit in preventing moisture migration, compared to the vertical unit, to the cold room. The room sizes and ceiling heights used seemed to have a negative effect on the measured open doorway infiltration results due to recirculation of air back through the opening. In typical cold storage applications where warm and cold room sizes may be as large as 200,000 square feet and ceiling heights at 30-ft the recirculation of air through the opening would not exist. In these applications we would conclude that horizontal recirculatory air curtains in this same temperature range would have effectiveness values as high as 80% and the vertical non-recirculatory air curtain would have effectiveness values as high as 56%. The recirculatory air curtain value compares very well to the theoretical calculated value of 80.6%.

At 80% effectiveness the horizontal recirculatory air curtain would be compared to a typical fast acting traffic door operating at 20% door open time (DOT). The issue with the traffic door is that it provides no barrier when the door is open and this can create significant energy issues in extremely high traffic applications. In high traffic applications the single horizontal recirculatory air curtain would save energy when standard traffic doors are operating above a 20% DOT. This very high traffic exists in many large grocery distribution centers. Many other cold storage facilities; however, do not have this extremely high traffic. In these applications the coupling of a horizontal recirculatory air curtains with a fast acting traffic door will provide the highest energy savings possible. If an 80% effective barrier is provided whenever the traffic door is open and the traffic door provides a 95% + effective barrier whenever it is closed. The net result is extremely low energy loss at a doorway.

Future Work

This experiment focused on application of a horizontal recirculatory air curtain as a thermal barrier in a static cooler application. Future work would be to test double air vestibules in static freezer applications and expand the single and double testing to non-static pressure imbalanced doorways. If larger calorimeter rooms become available it would also be a benefit to test the conclusion of open doorway infiltration numbers being effected by room size and ceiling height.

Acknowledgements

All of the testing for this experiment was completed by Creative thermal Solutions of Urbana, IL. They provided a very comprehensive thermal sciences laboratory that met the needs of this type of testing. Special thanks to Dr. Pega Hrnjak, Phd., Sefan Elbel, Phd., and Andy Musser for their hard work and providing the facility and calorimeter room.

References

1. ASHRAE. 1990. 1990 ASHRAE handbook-Refrigeration systems and applications. Atlanta: American Society of Heating, Refrigerating and Air- Conditioning Engineers, Inc.
2. ASHRAE. 2002. 2002 ASHRAE handbook-Refrigeration systems and applications. Atlanta: American Society of Heating, Refrigerating and Air- Conditioning Engineers, Inc.
3. Gosney, W.B., and H.A. Olama. 1975. Heat and enthalpy gains through cold room doorways. Paper presented before the Institute of Refrigeration at the Faculty of Environmental Science and Technology, The Polytechnic of South Bank, London.
4. Hendrix, W.A.,D.R. Henderson, and H.Z. Jackson. 1989. Infiltration heat gains through cold-storage room doorways. ASHRAE *Transactions* 95(2).
5. Longdill, G.R., and L.G. Wyborn. 1978. Performance of air curtain in single-story cold stores. *Proceedings of* 15th *International Congress of Refrigeration*, Venice, Vo. IV, pp. 77-88.
6. Pham, Q.T., and D.W. Oliver. 1983. Infiltration of air into cold stores. *Proceedings of* 16th *International Congress of Refrigeration*, Paris, Vol. IV, pp. 67-72.
7. Valkeapaa A., Siren K., Raapana I. 2005. Air Leakage Through Horizontal Air Curtain – An Experimental Study. *Finnish Institute of Occupational Health, Oulu, Finland, Helsinki University of Technology, Espoo, Finland.*

Appendix A

Model AC
Energy Aanlysis

PROJECT

CUSTOMER	Creative Thermal Solutions	DATE	11/16/10
DOOR DESCRIPTION	7-ft by 7-ft Doorway	PROJ #	0

ENGINEERING DATA

DOORWAY SIZE			OPERATIONS		
WIDTH	84	inches	ANNUAL WAREHOUSING HRS	5840	hrs
HEIGHT	84	inches	UNIT COST OF ELECTRICITY	0.10	$/kWh
CONDITIONS					
FREEZER TEMPERATURE	39	°F	ANTEROOM TEMPERATURE	75	°F
RELATIVE HUMIDITY	80	%	RELATIVE HUMIDITY	60	%
AIR-CURTAIN PROPERTIES					
AC-1 TEMPERATURE	36.4	°F	AC-2 TEMPERATURE	N/A	°F
UNIT HEATER	N/A	kw	AC-3 TEMPERATURE	N/A	°F
CONTRIBUTION TO FREEZER LOAD			**BENEFIT TO ANTEROOM**		
SENSIBLE HEAT	1.50	ton	SENSIBLE	-1.5	ton
LATENT HEAT	1.32	ton	LATENT	-1.3	ton
EQUIPMENT RELATED LOAD					
SENSIBLE HEAT		ton	Exfiltration of the lower humidity air-curtain normally provides a latent benefit to anteroom conditions.		
LATENT HEAT		ton			
TOTAL	2.82	ton	TOTAL	-2.8	ton

TROPICANA'S OPTIMIZATION OF LANDFILL METHANE GAS USE

Glenn Johnson, Senior Principal Sustainability Engineer
Tropicana Manufacturing Company Inc.
1001 13th Avenue East
Bradenton, FL 34208
Email: Glenn.Johnson@pepsico.com

Biographical Sketch

Mr. Johnson identifies, develops, justifies, and executes sustainability projects for Pepsi Americas Beverages. Prior to the sustainability role he served as Senior Manager of Automation and Utilities for Tropicana Corporate Engineering. He has a BSEE from the University of Illinois Urbana and a n MBA from California State University Fullerton. During his career he has justified and managed numerous cogeneration, refrigeration, thermal efficiency, power distribution, compressed air, and water projects and serves as a subject matter expert to PepsiCo's Global Engineering Team in a number of these areas.

Project Abstract

In 2005, Tropicana's Fort Pierce, Florida manufacturing facility entered into a long term agreement to purchase the landfill methane gas (LMG) from the Saint Lucie County Solid Waste Baling and Recycling center which shares Tropicana's northern border. From 2005 through 2011, Tropicana used the LMG to fuel a converted, 50,000 lbs./hr. water-tube boiler. In 2011, as that boiler started to fail and LMG production was expected to grow, Tropicana increased its investment in LMG. To fully utilize the anticipated volume of LMG, an additional boiler was converted for LMG use and a 1600 kilowatt, LMG-fueled, cogeneration facility was installed that incorporates a Caterpillar 3250C engine and includes both a Heat Recovery Steam Generator (HRSG) and a jacket water heat recovery system. All of these projects have provided significant environmental and financial benefits to both PepsiCo and Saint Lucie County. This presentation will provide an historical perspective of the project development logic, execution, and results to date.

TO TREAT OR NOT TO TREAT CITRUS WASTEWATER: WHAT ARE THE ALTERNATIVES?

Denise Johnston, Vice President Commercial, Veolia Water Solutions & Technologies
6981 N. Park Drive, Suite 600
Pennsauken, NJ 08109
Email: denise.johnston@veolia.com

From being just one more utility, wastewater treatment has become a potential source of energy, water reuse, by product recovery, regulatory compliance and good corporate citizenship. From conventional solids separation, pH adjustment and aerobic treatment, to anaerobic digestion with energy recovery to the use of membranes for final purification, the spectrum of wastewater treatment solutions has become a very interesting topic with numerous alternatives. This presentation discusses several of the main treatment alternatives.

TO TREAT OR NOT TO TREAT CITRUS WASTEWATER: WHAT ARE THE ALTERNATIVES?

Jill A. Jordan
jill.jordan@veolia.com

Denise Johnston, Vice President Commercial, Veolia Water Solutions & Technologies
6981 N. Park Drive, Suite 600
Pennsauken, NJ 08109
Email: denise.johnston@veolia.com

ABSTRACT

From being just one more utility, wastewater treatment has become a potential source of energy, water reuse, by product recovery, regulatory compliance and good corporate citizenship. From conventional solids separation, pH adjustment and aerobic treatment, to anaerobic digestion with energy recovery to the use of membranes for final purification, the spectrum of wastewater treatment solutions has become a very interesting topic with numerous alternatives. This paper discusses several of the main treatment alternatives.

INTRODUCTION

This presentation today is intended to familiarize you with the various alternatives that are available to treat wastewater generated at Citrus facilities. It will be concentrated on biological treatment more than physical / chemical alternatives. Conventional wastewater treatment utilizes predominantly aerobic biological processes, which, due to the aeration requirements, consume large amounts of electrical energy. In the past ten (10) to fifteen years (15), anaerobic treatment has found a place in treatment of industrial wastewater, requiring less energy and footprint as compared to aerobic treatment.

BIOLOGICAL WASTEWATER TREATMENT

Biological wastewater treatment generally includes the following unit process:

1. A pre-treatment step. This may include clarification, equalization, detoxification neutralization etc. to prepare the wastewater for biological treatment;
2. The actual biological step. Biomass is brought to intimate contact with the wastewater under controlled conditions and degradation of waste (organic) material is taking place;
3. A separation or clarification step to separate biomass from treated wastewater; and
4. A polishing step. This is not required at every site depending on the specific discharge requirements.

AEROBIC AND ANAEROBIC ALTERNATIVES

Wastewater is typically treated aerobically or anaerobically with the major difference that aerobic micro-organisms require oxygen and anaerobic organisms do not need or like oxygen.

There are a variety of aerobic and anaerobic technologies available that have been proven to be successful in the treatment of Food and Beverage, and in particular Citrus Wastewater. A Conventional Activated Sludge (aerobic) system is provided in a few different configurations. There is the well know concentric system which is provided as a "tank within a tank". The outside of the tank is where the aeration takes place, and the inside of the tank is where the clarification is accomplished. Another configuration that most people are familiar with is the large above or below ground basin. Several different aerobic schemes are employed; such as floating aerators or "snake" line aeration. Coarse and Fine bubble diffusion is available, depending on the influent characteristics of the wastewater. Gaining more popularity is the MBBR aerobic system (Moving Bed Bio-Reactor) which is filled with media and is able to operate at higher loadings than conventional aerobic.

There are also a variety of anaerobic technologies available and are typically categorized by low rate, medium rate and high rate. The difference from one anaerobic technology to another is how the sludge is kept in the reactor to ensure good sludge to wastewater contact. A low rate system is well recognized as an in ground lagoon. The medium rate system is recognized as a CSTR (continuously stirred tank reactor) and is designed to degrade high solids and FOG's in a stream. The high rate reactors are the most popular worldwide with more installations in the Food and Beverage industry than any other type of anaerobic technology.

When comparing the two technologies, there are several decision making factors that are typically considered. The Table below gives a quick overview of particulars that a Citrus Plant would consider:

Parameter	Aerobic	Anaerobic
Achieves <100 mg/l	Yes	No - needs aerobic polishing
Can be shutdown	No	Yes - for extended periods
Sludge Generation	High	Low
Energy Usage	High	Low
Land Usage	High	Low
Useable Biogas	No	Yes

Typically, anaerobic technology is favored over aerobic technology when the following considerations are evaluated:

1. Treatment of high strength wastewater. An influent BOD concentration of >1100 mg/l;
2. Space Constrained
3. High Operating Cost with the present system, such as surcharges, and energy usage;
4. A need for shutdown capability for a period of a week or more; and
5. An internal outlet for the biogas that is produced such as the production plant boiler a use for electricity from co-generation.

Numerous facilities throughout the world have employed anaerobic treatment followed by aerobic polishing. The combination of technologies results in

1. Aerobic polishing step is smaller than straight aerobic treatment of the entire stream due to the anaerobic pre-treatment step removing 85% - 95% of the influent load;
2. The reduced size of the aerobic polishing step results in a significant reduction in energy usage;
3. Less sludge generation due to the downsizing of the aerobic system; and
4. Acute toxicity requirements are achieved.

ANAEROBIC DESIGN PARAMETERS

As discussed earlier, there are a handful of popular anaerobic technologies, and they are typically characterized by what the hydraulic retention time is within the reactor. All technologies mentioned below most usually designed to operate in the mesophyllic (35°C). Thermophyllic reactors are in operation, but in much smaller number.

Low Rate System – The Volumetric Load associated with this type of system is 0.5 – 2.0 kgCOD/m3/d, with a hydraulic retention time typically 10-25 days depending upon the wastewater. This type of system is ideal for high solids, high FOG wastewater. It is characterized by taking a very large land area and can be in or above ground. The reactor is intermittently mixed.

Medium Rate System – There are two popular types of Medium Rate Systems – those being CSTR and Anaerobic MBR. In both instances, a mechanically mixed reactor is utilized. The CSTR system may be a "once through" reactor, but more common in the food industry is a CSTR reactor with solids separation (clarifier or DAF) on the back end where sludge is recycled back into the digestion vessel. When coupled with solids separation, the hydraulic retention time is typically 3 – 7 days. In a once through reactor (popular in the manure digestion arena) the hydraulic retention time is 20 – 25 days. The medium rate reactor is again ideal for high concentration (>30,000 mg/l COD), high solids and high FOG waste streams.

High Rate System – This system is designed to treat the soluble portion of the waste streams, as the hydraulic retention time is measured in hours (3 – 72 hrs) rather than days. Used extensively in the Food and Beverage industries, this is the most compact anaerobic reactor system. High rate systems use granular sludge which is brought in from operating systems, which ensures a quick start up.

In an anaerobic reactor, there are two basic types of bacteria, acid forming and methane forming. The Acid forming bacteria can live in a wide range of pH and temperature conditions. They reproduce every 3 – 5 minutes and must be controlled not to outgrow the methane forming bacteria. The job of the Acid Formers is to convert carbohydrates into short chain acids that the Methane Formers eat.

The Methane Forming Bacteria work optimally in a narrow pH range (6.6 - 7.6 s.u.) as well as in a narrow temperature range (85 - 100°F). They reproduce every 3 – 5 days and are responsible for converting the acids into CH_4 and CO_2 (Biogas).
The key to process stability is to recognize that the methane formation is the rate limiting step in most anaerobic applications. One must keep the production rate of Volatile Fatty Acids or VFA at or below the consumption rate of VFA. Low effluent VFA (<5meq/l) signals a healthy system operation.

The Operational Parameters associated with an anaerobic reactor are:

- pH – 6.7 – 7.6 s.u.
- Temperature 85° - 100°F
- COD:N:P – 500:5:1 (for COD <5000 mg/l)
- COD:N:P – 850:5:1 (for COD >5000 mg/l)

Outside the parameters given above, the reactor may operate satisfactorily, but biomass quality can be compromised. Remember.....Bugs are people too! They eat, digest, make gas, reproduce and die. People thrive in a healthy environment.....so does sludge.

The treatment scheme for treating Citrus Wastewater will typically have Grit Removal as the first step, equalization (unless feeding to a low rate system), DAF if there is a d-limonene concern, then anaerobic digestion. Post treatment schemes vary on the effluent criteria that one must meet.

If the anaerobic system is being utilized as pre-treatment to POTW surcharge, then typically there is no need for post treatment. An anaerobic system will reliably achieve BOD removal efficiencies of 90% or more treating a Citrus Wastewater. Biogas generation is based upon the theoretical value of 0.35 m3 CH_4/kg COD removed. The biogas composition in a Citrus application will contain a CH_4 of 75%, CO_2 of 20-23%, it will be saturated and also have negligible amounts of other compounds.

If the anaerobic system is being utilized as an NPDES discharger, then a polishing step would be coupled with the reactor. The polishing step may be in the form of membranes (as in the AnMBR), or an aerobic system. The combination of anaerobic plus aerobic polishing may be designed to achieve effluent limits of 25/25 if appropriate.

AEROBIC DESIGN PARAMETERS

Aerobic technologies have been widely used to treat both industrial and municipal wastewater for many years. Solids Retention Time (SRT) and oxygen transfer are the two main parameters for aerobic processes design:

- SRT determines the biological ecology. Most process engineers calculate the tank volume using SRT, V ———, where WAS is the sludge wasting rate that can be can be calculated based on cell synthesis, particulate COD hydrolysis, inorganic solids precipitation and endogenous cell decay.
- After knowing how much COD will grow into biomass and how much COD in the effluent, we can calculate how much COD will be bio-degraded using oxygen. That is called Actual Oxygen Requirement (AOR). Then we can calculate the aeration system based on the oxygen transfer efficiency of the fine bubble or coarse bubble system, mixed liquor concentration, water depth, temperature and expected residue

dissolved oxygen concentration. Eventually the blower, the layout of the air diffusers and the instrument & control can be designed.

Besides removing COD, the aerobic process can also do nitrification, de-nitrification and enhanced biological phosphorous removal. Nitrification needs long SRT and oxygen to convert ammonia into nitrate. De-nitrification takes place in an anoxic zone, where nitrate accepts electrons from COD or ammonia so nitrogen is eventually removed mainly as nitrogen gas. Enhanced biological phosphorus removal process needs an alternated anaerobic and aerobic environment, so that the phosphorus accumulating organisms can out compete with normal heterotrophic, and phosphorus eventually removed in the biological sludge.

Aerobic processes, with the proper SRT, normally have good Extracellular Polymeric Substances (EPS) structure. That is why aerobic sludge is a good bio-flocculant to capture the colloidal COD in the wastewater, and aerobic sludge normally settles pretty well.
In summary, effluent of the aerobic process will have low COD, nitrogen, phosphorus and TSS.

TREATING CITRUS WASTEWATER

In 1985 an anaerobic pilot study was conducted at a U.S. Pectin production facility. The pectin was produced from Citrus Peel. This was the first undertaking of anaerobic treatment for this type of wastewater. The effluent from the plant would first undergo treatment by a DAF (Dissolved Air Flotation) system to remove excessive solids and the effluent would then be forwarded to an anaerobic reactor. This study took place over a period of five (5) months and proved to be successful enough to continue with a full scale treatment system. BOD removals at this plant were consistently in the 85% - 90% range. Biogas, with an average methane purity of 75% was used to offset natural gas usage at the plant.

In 1991 an anaerobic full scale plant was installed for a Citrus production facility in Israel. Once installed, problems were encountered, and after extensive toxicity tests, it was determined that the concentration of d-limonene in the wastewater was above toxic levels. At this particular production facility, it was decided that the stream containing the highest concentration of d-limonene would bypass the anaerobic process and be directed straight to the aerobic process.
Presently, d-limonene is thought to be inhibitory at a concentration of 20-70 mg/l, and toxic at a concentration of >70 mg/l.

Subsequent to the two anaerobic experiences provided above, work was undertaken to determine the threshold of d-limonene acceptable to anaerobic technology. Further, it was determined that should a stream contain more than treatable anaerobically, and upfront step such as a DAF would strip this compound and the wastewater would then be treatable anaerobically.

The Tropicana installation in Bradenton first installed a media or filter type anaerobic reactor and have recently had it retro-fitted to a UASB type of reactor. This application is presently achieving an SCOD removal efficiency of more than 90%.

Approximately five (5) years ago, a Citrus Plant in Florida evaluated the costs associated with the existing aerobic treatment plant to determine if they could add improvements to lower the operating costs associated with the treatment scheme. It was determined that installation of an anaerobic system preceding the aerobic system would result in a significant reduction in sludge generation (approximately 55% less per year), the electrical usage would be reduced by approximately 60% per year and the production of usable biogas would be on the order of 65,000 MM Btu/yr.

SUMMARY

Based on the extensive worldwide experience treating Citrus wastewater biologically, it is fair to state that it is an excellent substrate for anaerobic and/or aerobic treatment. Consideration of the site specific characteristics and economics will help drive the Citrus plant to the treatment system that best fits their needs. If using anaerobic technology, one must test for the d-limonene concentration in the influent.

Anaerobic treatment is an alternative to conventional aerobic treatment due to its lower solids generation rate, low electrical energy requirements, and the production of a usable biogas which is, of course, renewable energy. The coupling of anaerobic as a 1st stage with aerobic treatment as a 2nd stage provides a very high level of wastewater treatment with lower energy requirements and lower sludge production than a conventional aerobic activated sludge plant.

GREENING PANEL DISCUSSION
OUR FUTURE WITH HLB: WHERE TO FROM HERE?

Moderator

Dan King, Director of Scientific Research, Florida Department of Citrus

Panelists

Elizabeth A. Baldwin, Supervisory Research Horticulturalist, USDA-ARS
Mike Irey, Director of Research, US Sugar/Southern Gardens Citrus
Dana D. Stump, President, Direct Driven Solutions LLC
Jim Graham, Professor of Soil Microbiology, UF/IFAS CREC
Greg McCollum, Research Physiologist, USDA-ARS

CEC TRANSACTION AND PRESENTATION TITLES SINCE 1955

Volume I – 1955

Problems Facing the Citrus Industry	H.W. Schwarz
Water Treatment in Citrus Processing Plants	A.T. Lohkamp
Precooling of Citrus Fruits	G.E. Sutton
Design Considerations for Cold Storage Warehousing	Jim Fiske
Progress in Waste Treatment in the Citrus Industry	John Wakefield
Corrosion Resistant Materials for the Citrus Processing Industry	W.T. Tiffin

Volume II - 1956

Factors Affecting Selection of Refrigeration Equipment in the Citrus Industry	Harland Rex, Sr.
Spray Irrigation of Citrus Waste	W.H. Young
Problems Relating to Production and Distribution of Cartoned Orange Juice	David O. Hamrick
Products from Citrus Peel	R.W. Kilburn
Getting the Most from your Evaporator	E.J. Kelly
Handling and Packing Trends in Fresh Fruit Operations	T.R. Perry

Volume III - 1957

Can Handling Systems	J.F. Stilling
Booster Compressors for the Citrus Industry-Reciprocating, Rotary and Centrifugal	L.R. St. Onge, P.E.
Steam Generation for the Citrus Industry	Roger Holmes
Production and Distribution of Chilled Citrus Salads	David O. Hamrick
Continuous Vacuum Dehydration of Citrus Juices	J.M. Bonnell
Modern Handling of Citrus Fruit from Grove to Plant	Harry A. Clapp, E.J. Dunham
Cost Advantages of Bulk Handling Through the Citrus Packing House	R.V. Phillips

Volume IV - 1958

Automatic Palletizing	G.D. Beaver
Points to Consider in Using Natural Gas as a Fuel in Citrus Plants	J.W. Humphrey
Treating and Mechanical Packaging for the Fresh Fruit Packing House	Samuel Ellison
Automation for the Citrus Industry	Jack Barker, P.E.
Some Practical Aspects of Basic Hydraulics	Courtney A. Forth
In-Place Cleaning of Concentrate Plant Equipment with Special Emphasis on Extractors and Evaporators	D.I. Murdock
Some Engineering Aspects of New Metal Container Materials	Dr. R.W. Pilcher

Volume V - 1959

Electronic Brix Indication and Control as Applied to the Citrus Concentrate Industry	Francis S. Reed
Factors Involved in Selecting Optimum Size Industrial Valves & Pipes	Earl F. Riopelle
The Use and Results of the Wind Machine to Prevent Freeze Damage	J.G. Georg
Steam Economics of Molasses Plant Evaporators	John M. Cahill

Electrical Design & Application Fundamentals for the Citrus Industry	James I. Gay
Improved Equipment & Methods for Harvesting Citrus	G.E. Coppock
Water Extraction of Fruit Solids from Orange Pulp	F.W. Wenzel, R.L. Huggart, R.W. Olsen

Volume VI - 1960

Old & New Ideas in the Citrus Pulp Industry	D.S. Prosser, Jr.
Automatic Lubrication	Herbert J. Brouillette
Process Treatment of Citrus Fruit for the Fresh Fruit Markets	R.D. Gerwe
Electric Insulating for the Citrus Industry	L. Ed Mulder
Electric Powered vs. Gas Powered Lift Truck	G.A. Markell
Stainless Steels for the Citrus Processing Industry	William T. Tiffin
Nitrogen Treatment as a Means of Extending Shelf-Life of Chilled and Canned Single Strength Orange Juice	Robert G. Webster

Volume VII - 1961

Aluminum Cans for the Citrus Industry	M. Neil Burleson
A Discussion of the Methods for the Production of Essential Oils in Florida	James W. Kesterson
Treatment of Citrus Juice by Ion Exchange Membrane	Robert W. Kilburn
A Brix-o-Meter for Fresh Juice Testing	Francis S. Reed
Flavor Retention of Orange Juice	Thurston W. Harwell
A.P.V. Plate Evaporator in the Citrus Industry	D.F. Dinnage
Alcohol & Glycol Type Brines vs. Air Blast Freezing	A.T. Lohkamp

Volume VIII - 1962

Treatment & Disposal of Citrus Processing Waste Water	Dr. R.R. McNary
Can Handling Principles as Applied to Today's 6 oz. Concentrate Can	D.O. Bair
Centrifuges in the Citrus Industry	Adolf Distelkamp
Use of Fluid Mixers in the Citrus Industry	J.Y. Oldshue
Packaging for the Citrus Industry	A.H. Lucas, Jr.
Progress of Foam-Mat Drying of Orange Juice	M.K. Veldhuis, J.H. Tatus, C.J. Wagner, Jr., O.W. Bissett
Multiple Packaging	Frank K. Lawler

Volume IX - 1963

High Temperature Short Time Evaporation	Ralph W. Cook, P.E.
Sargeant Electronic Evaporator	William R. Erickson
Modern Fresh Fruit Handling, Adams Packing Assoc., Inc.	W. Grierson
Engineering Design & Economics of Air Conditioning in Florida's Newest Sectionizing Plant	William S. Ostrander
Automatic Lubrication	T.M. Murphy
Automated Control of Fruit Flow & Pump Washing	W.R. MacClarence, A.A. Lang
Rough Handling of Filled Cans as Related to Leakage Type Spoilage	V.S. Troy

Volume X - 1964

Pulp Wash System Development	Ronald B. McKinnis, Richard A. Andrews, Herman L. Jones
Pulp Washing with FMC Inline Extractors	Wilber C. Belk
Integral Rivet Easy Open Ends	Phillip C. Althen
Application of Centrifuges in the Recovery of Citrus Flavor Fractions	R.E. Thrush
Fundamentals of Heat Transfer Applicable to the Citrus Industry	Dr. Erich A. Farber
Fundamentals of Glass Handling	C.P. Whittier
Research to Improve Methods for the Determination of Pounds-Solids	James G. Blair

Volume XI - 1965

The Market Impact of Synthetics	Edward A. Taylor
Pressure Control on Carton & Pallet Accumulation Conveyors, Bulk Container Conveyor Systems	George J. Dawson
Improved Maintenance at Reduced Costs	Kenneth S. Caldwell
Composite Cans for Frozen Citrus Concentrate Products	William E. Hughes
Economic Justification of Capital Expenditures	J.E. Deuel
A Procedure for Determination of Fruit Quality at Time of Receipt	B.M. Middlebrooks
Citrus Processing Outside of the United States	F.E. Penn

Volume XII - 1966

Wrap Around Case Packing	Walter D. Wood
Dynamics of Agriculture	Dr. Ken R. Tefertiller
Liquid Nitrogen Refrigeration	David O. Jones
Fundamentals of Fluid Flow Applicable to the Citrus Industry	Erich A. Farber
Cold Protection of Citrus with Petroleum Coke, Wax or Liquefied Gas	Dr. John F. Gerber
Solid State Controls for Industrial Trucks	Richard Towne
The Changing Fruit Beverage Market	F.G. Berner

Volume XIII - 1967

Interpretation of Present Florida Pollution Control Regulations	A.T. Nielsen
Data Processing from the Field to the Super-Market	Edward Hammerle
Unitized Distribution Systems	Lawrence F. Swanson
Waste Heat Recovery from Feed Mill Dryers	R. Paul Snyder, P.E.
A Basic View of Solid State Electronics	Dr. Bruce E. Mathews
Canned Chilled Juice	John M. Boyd
Storage and Handling of Dry Citrus Pulp	W.S. Boots

Volume XIV - 1968

Measuring Stack Particulate Emission	John W. Seabury, P.E.
Considerations for Air Power Systems	Robert S. Johnson
Fluidics - A New Tool for Controls	Dr. Erich A. Farber
A New Approach to Evaporation	Derek F. Dinnage
A Non-Metallic Answer to Corrosion Problems	Robert R. Willis, Eugene F. Jacobs

Recovered Volatiles from Citrus Juices	R.W Wolford, Dr. L.G. MacDowell, C.D. Atkins, M.H. Dougherty
Bulk Storage of Dried Citrus Pulp	William Hugh Young

Volume XV - 1969

Relationships and Legal Aspects of Florida State Board of Health and Florida Air and Water Pollution Control Commission	Nick Mastro
The Use of Nitrogen in Freezing Citrus Concentrates	Paul Smith
Packing House Automation in California	Maurice Johnson
Air Pollution Abatement Equipment and Processes	E.R. Hendrickson, Jorgen G. Hedenhag
Legal Requirement and Liabilities of Canner as Well as Equipment Supplier	Donald B. Wilcox
Uses of Statistics in Citrus Operations	William J. Hepburn
Can Making Technology	Ronald H. Armbruster

Volume XVI - 1970

Mid-Peninsula Future Water Requirements, Uses, Priorities, and Economics	Dr. R.O. Vernon
Hydraulic Conveyance of Citrus Fruits	Mr. Aaron Curls
A Comparison of Iodophor and Sodium Hypochlorite Sanitizers	R.B. Barrett
Manufacture of Spiral Wound Composite Cans	Roy M. Martin
Manufacture of Plastic Containers for Citrus	Elwyn Jones
Vibration and Noise	Dr. Erich Farber
Citrus Pectin and Dried Pectin Peel	Paul Snyder

Volume XVII - 1971

Citrus Pulp Pelleting	William G. Trapp
Control of Noise Problems in Citrus Plants	Dr. Calvin C. Oliver
Shrink Film - Manufacture, Properties, Application	E.E. Quigley
Is Your Evaporator Polluting	Harold Rebeck
Imitation Flavoring from Citrus Oils	C.E. Desch
Factors Affecting Electric Motor Selection	Prof. Edward I. Smith
Nutrients: A Growing Problem in Pollution Control	Vincent D. Patten

Volume XVIII - 1972

d'Limonene Recovery	H.E. Schulz
Process Computers in Citrus	R.E.J. Putnam, Murray D. McMahon
Sugar Handling Systems	Walter Kress
Legal Aspects of Air and Water Pollution	Kenneth F. Hoffman
Floor Coverings for Citrus Plants	John A. Wasser
Occupational Safety and Health Act of 1970	Mr. Theodore F. Schoenborn, Dr. Erich Farber
Tank Storage of Bulk Concentrate	R.A. Riley

Volume XIX - 1973	
Energy - Resources & Utilization	Professor Erich A. Farber
Are you Abusing your Boiler?	Ken Robinson, Bill Segrist
Practical Noise Control	Avery L. Owen
Cooling Towers for Process Industries	Max Bell
The Use of Fabric Filters in Air Pollution Control	Tom Brennan
Potentials for Photoelectric Grading Equipment	J.J. Gaffney
High Efficiency Can Lines	G.E. Van Keuren
Volume XX - 1974	
Bulk Storage of Citrus Juice Concentrate	Morris W. Ratcliff
Spray Irrigation of Citrus Waste	Henry P. Morrell
Progress on Mechanically Grading Citrus Fruit in Processing Plants	Dr. William L. Bryan
The Food Outlook: A World Imperative	Dr. Ralph A. Eastwood
Economics of d-Limonene Recovery	Gordon P. Gerow
Centralized Control for the Citrus Concentrating Plant	Ed Stephenson
The Fuel Crisis	W.J. Smith
Volume XXI - 1975	
Potable Water from Plant Waste Water	Gordon H. Richardson, Omer R. McDuff
Increasing Boiler Efficiencies	Richard P. O'Leary
Handling Lemons in a Conventional Packing House	G. Martin Stephens
Availability and Economic Possibilities of Fuels	Dr. Erich A. Farber
Supplemental Power Generation	Richard L. Chamberland
Economics of Citrus Specialty Products	Gordon P. Gerow, L.E. Grey
Technological Change in the Food Industry	Mark C. Hollis
Volume XXII - 1976	
Spray Irrigation of Waste Water on Citrus Groves	R.C.J. Koo
Condensate Reuse in Citrus Plants	David O. Hamrick
The Economics of Power Factor Correction	T.I. Wetherington
Comparison and Application of Refrigeration Compressors	Hunter Terrett, Larry Kelly
The Recovery of Sugar from Citrus Press Liquor	B.R. Breslau, D.F. Pensenstadler, W.G. Mitchell
Trash and Stem Removal from Incoming Fruit	R. Paul Snyder
Florida Citrus Industry and the Engineer	Thos. B. Kimball
Volume XXIII - 1977	
The Metric System as it Applies to the Citrus Industry	R.C. Rapson, Jr.
Solid State A-C Variable Speed Drives	John D. Zak
Multiple Hydraulic Drives Load Sensing	W.C. Adams, Chuck Mueller
Citrus in the World Food Outlook	Ralph A. Eastwood
Citrus in Brazil	Peter C. Mayer

Water - The Essence of Life	G. Lionel Bailey
Condensate Reuse - Phase II	Richard A. Hamilton
Energy Savings in the Citrus Industry	Michael W. Graybill

Volume XXIV - 1978

Coal as a Fuel	Alex Kaiser
Irrigation Methods & Equipment	Dalton S. Harrison
Industrial Control as Programmable Controllers	Victor W. Reed
Dietary Citrus Fibers	R.R. Ferguson, Dr. K.I. Fox
Recovery of Citrus By-Products & Specialty Products	James W. Kesterson
Energy Review	Eric Zausner
Pumping Viscous Products	Clarence L. Kelley

Volume XXV - 1979

Problems Facing the Citrus Industry - A Reprise	H.W. Schwarz
Florida's Revised Water Quality Regulations	John C. Bottcher
Solar Energy: Feasibility for the Citrus Industry	A.T. Lohkamp
Peel Oil Recovery by Recycling Centrifuge Effluent	Dr. Elizabeth S. Steger
Weather Analysis and Forecasting	Frederick L. Crosby
Chlorine Dioxide - An Effective Biocide for Recycled or Reused Water Systems	John F. Synan
Cogeneration: An Example Within the Citrus Industry	Dennis D. Weimer

Volume XXVI - 1980

What Directions Next for the Citrus Industry	H.W. Schwarz
What's in a Citrus Fruit?	J.W. Kesterson
Conservation Begins in the Boiler Room	George G. Schrader
Coal-Oil Mixtures for Industrial Boilers	Jon Wilda, Joe Cochran
The Economics of Using Mechanical Vapor Recompression Evaporators to Concentrate Orange Juice	Larry R. Kelso, Charles M. Rowan, Jr., Kenneth L. Holladay
Deep-Well Injection in the Peninsula of Florida	Dr. J.I. Garcia-Bengochea, P.E.
Marketing Strategies for the 80's	Douglas R. Hoffer

Volume XXVII - 1981

Results of Energy Audits Conducted in Various Citrus Plans During 1979-80	Kenneth L. Holladay
Assessing Microcomputer Applications	J. Anthony Llewellyn
Fluidized Bed Combustion Packaged Boilers	John W. Wallish
Diesel Cogeneration	John Dunlay
Microcomputer Control of Commercial Citrus Taste Evaporators	C.S. Chen, R.D. Carter, C.J. Deimling, E.S. Moisa
Automated Juice Room Control	R.C. Stanley

Volume XXVIII - 1982

Preventive Maintenance with Infrared Detectors	Frank T. Caruso
Aseptic Packaging - State of the Art	Joe E. Marcy
Man/Machine Interface	Dr. Brian Murphy
Automated Pounds Solids Determination	Gordon P. Gerow
Centrifugal Pump Applications	James H. Poore
Low Level Heat Recovery with Heat Pumps	Joel S. Gilbert

Volume XXIX - 1983

Multiple Stage Pressing	Ralph W. Cook
The Information Explosion ACS STAC The Gateway to Information	Al Pozofsky, David R. Jenkins
Saving Energy on an Installed Ammonia System	Harold R. Heath
Environmental Law for Citrus Processors	Roger W. Schwenke
An Untapped Management Resource	William E. Hoffman, Paul Matchett
Florida Citrus Planning for the Future	Dan L. Gunter

Volume XXX - 1984

Freeze Concentration: An Alternative for Evaporation in the Citrus Industry	Ir. W.H.J.M. van Pelt, Ing. W.J.M. Swinkels
Economics of Peak Shaving and Cogeneration Using On-Site Generating Systems in the Citrus Industry	Dennis R. "Zeke" Sickinger
Wastewater Treatment Alternatives for a Food Plant	Alan Overcash
Considerations in the Manufacture of Alcohol from Citrus	James S. Hammond
Problems and Opportunities in the Use of Reconditioned Steel Drums	J.M. Murphy
Citrus Research in Florida: Issues, Policies, and Solutions	Walter J. Kender

Volume XXXI - 1985

Florida Citrus Freeze Losses and Recovery in the Field	Jodie D. Whitney
Insulation Thickness Guidelines for Energy Control, Process Control, Personnel Protection and Condensation Control	Winfield T. Irwin
Coal Rediscovered	Jon C. Wilda, Mark C. Elizer
Orange Juice Demand - An Industry in Transition	Dan L. Gunter
Cogeneration Systems and the Utility Interface	Thomas I. Wetherington
Fruit Handling Trucks to Juice Room	Palmer Tuthill

Volume XXXII - 1986

Cogeneration for the Simultaneous Supply of Power, Heat and Refrigeration	G. Holldorff, W. Malewski
Microcomputers in Citrus Plants	Richard L. Westmoreland
Automated Storage and Retrieval Systems and the Citrus Industry	Robert E. Gulardo
Citrus Canker: Impacts of Research on Eradication and Control	Walter J. Kender
Centrifuging Citrus Waste and Juice	John V. McKenna
FCOJ Bulk Transportation	C. Boyde Wynne, III

Volume XXXIII - 1987

Citrus Pulp Recovery	Teiko M. Johnson
Cogeneration and the Citrus Industry	Edwin F. Coxe
An Improved Alternating Current Motor	John F. Roesel, Jr.
Flow and Tank Inventory Measurement	Carlos A. Smith, Wayne Ralph
Automated Inspection/Classification of Fruits and Vegetables	Dr. William M. Miller
Citrus Peel Dryer Control System	Ralph W. Cook

Volume XXXIV - 1988

Industrial Refrigeration Design Safety	Henry B. Bonar, II
Caer. Sara and the Citrus Processor	Kenneth Ballard, Donald Wise
Environmental Permitting and Development of Citrus Groves in Southwest Florida	William M. Helfferich
Overview of Toxic Contaminants in Ground Water Origin, Transport, and Treatment	Brannon H. Wilder
The Schmidt Sigmaster Evaporator: An Improved Evaporator Type for the Fruit Juice Industry	Richard A. Aust
The Volatile Flavors of Orange Juice	James B. Redd

Volume XXXV - 1989

Citrus Irrigation with Reclaimed Wastewater	Brian R. McMahon, Dr. Robert C.J. Koo, H. Williams Persons, P.E.
Cogeneration - Theory and Practice	Richard D. Kennedy
Aseptic Bulk Handling of Citrus Juices and Concentrates	Glenna Matthews
Membrane Concentration of Orange Juice	Stephen Cross
Heat Exchanger Application and Theory	James G. Meholick
The T.A.S.T.E. Evaporator, Continuing Progress in Evaporator Technology	Stanford Brett Welch

Volume XXXVI - 1990

A Commercial Citrus Debittering System	Seth I. Norman
Citrus Packaging and the Environment	John McDonald
The Anaerobic Treatment of Food and Citrus Processing Wastewaters	Dr. L. Michael Szendrey
Homogenizing Concentrate in a Juice Evaporator	Philip Grant
Limonene - A Versatile Chemical	Joseph J. McBride
Aseptic Processing and Handling of Citrus Juices	Michael J. Sanders

Volume XXXVII - 1991

Status Update of the Worldwide Citrus Industry	Kenneth Fox., Ph.D.
Supercritical Carbon Dioxide Applied to Citrus Processing	Murat O. Balaban, Ana G. Arreola
The Outlook for Electrical Power in Florida	J. Ken Wiley

Citrus Debittering with Ultrafiltration/Adsorption Combined Technology	Mike Wethern
Programmable Logic Controllers	Terence M. Sages
TQM: The Quality Imperative	William Smitley, Ph.D., Teri Yanovitch

Volume XXXVIII - 1992

Corrosion	Jeffrey Raasch
The Role of Biotechnology in the Development of Improved Citrus Scion and Rootstock Cultivars	Jude W. Grosser
Product and Ingredient Security	William S. DuBose
A New Concept of Evaporation Plants: Thermal Vapor Recompression for Reducing Costs and Improving Product Quality	Bernhard W.H. Wiegand
Radar Liquid Level Measurement	Mark D. England
Bar Code Production Inventory Systems: An Integral Solution	Chris Stanberry

Volume XXXIX - 1993

Tank Reaction System for Citrus Peel	Carlos E. Odio
Title I of the Americans with Disabilities Act & the CitrusIndustry: The Mystery Unraveled	John M. Breckenridge, Jr.
Coldpressed Citrus Oil Recovery	Roger Waters
Mass Flow Meters and the Citrus Industry	Avedis N. Zarikian
Ammonia Refrigeration & OSHA Process SafetyManagement Programs	Steven J. Tyler
Citrus Juice Processing in Brazil	Daniel S. Marques

Volume XL - 1994

New Generation of State Test House Equipment	Roy Ratajeski
Integrated Citrus Information System	Dennis Davis
Workplace Ergonomics: An Approach to Health, Safety, and Productivity	Gary Kress, Ph.D.
Process Safety Management An Outline for Compliance" "	Donald L. Julitz
Water Use and Wastewater Flow Reduction in the Citrus Processing Industry	Dr. Richard H. Jones, P.E., David A. Keough, P.E.
Energy Efficiency Opportunities for Citrus Processing	Jeannie Benson

Volume XLI - 1995

Variable Speed Drives	Harold T. Snyder, P.E.
Deregulation in the Electric Power Industry	Steven G. Rudolph
A Review of Citrus Harvesting in Florida	Jodie D. Whitney, Ph.D.
Latest Developments in Pressing Citrus Peel	Robert B. Johnston
Warehouse Configuration Analysis to Achieve Productivity and Cube Utilization	Richard F. Schieler
Short Time Reaction System for Improved Citrus Peel Processing	Daniel S. Marques

Volume XLII - 1996

Best Practices for Fresh Juice Production	Greg W. Schrader, Lisamarie Kane
DuPont: Safety Management in a Re-Engineered Corporate Culture	Thomas C. Lemm
Deoiling Single-Strength Orange Juice	Joseph A. Puglia Douglas P. Harper
d-Limonene Recovery	Carlos Odio
Asset Management Perception for Industrial Compressed Air Systems	Henry L. Kemp, Jr.
Aseptic Processing and Storage of Citrus Juices	Jeffrey B. Raasch

Volume XLIII - 1997

Feed Mill Automation (To Improve Efficiency and Quality)	Glenn Sykora
Thermal Imaging of Electrical and Mechanical Systems	Joe Symmes
How to Calculate Return on Investment for Maintenance Improvement Projects	Terry Wireman
Pumping Power Requirements and Role of Viscosity During Agitation of Concentrated Orange Juice	M.A. Rao
VOC Emissions from Citrus Processing Plants	David A. Buff, P.E.
The Internet and the Citrus Industry	Tony Montanari, Ph.D.

Volume XLIV - 1998

High Pressure Pasteurization of Citrus Juices	R.J. Braddock, M.E. Parish, J.K. Goodner
Pumps in Industry: Theory and Application	Walter Weaver, Eric Bradner, P.E.
When it Rains on the Citrus Industry – Regulatory Stormwater Agencies and their Requirements	Lori Webb-Paris, P.E.
Citrus Plant and Process Ventilation	Peter K. Ousley
Simulation Modeling: Risk Free Evaluation of Performance Alternatives	Dennis A. Schaffer
Pumped Peel - Five Years Later	Robert Johnston, P.E.

Volume XLV - 1999

New Burner Technologies	Doug Fillmon
Citrus Tank Farm Level Control	Doug Groh
Wide Belt Optical Inspection of Whole Citrus for Processing	Tim Reardon
A Call to Action: Citrus Industry Implications of the Year 2000 Embedded Systems Crisis	Rick Montney
HACCP for Citrus Processing	Debbie Newslow
Citrus Waste Water Technology	David Keough

Volume XLVI - 2000

Strategic Automation for Citrus Processing	David Borrison
Trends in Automation and Control	Doug Burns

Continuous Blending for Processes	Ed Ulvila
Designing a Sanitary Process System	Joe Pastore, APV
Sanitary Welding Standards	Richard E. Avery
CIP Systems - Design and Functions	Ashok Nangal, Ph.D.

Volume XLVII - 2001

Maintenance Best Practices	Lyle B. McCormick, Jr.
Ammonia Refrigeration Automation & Efficiencies	Dean Ellerbrock
Manufacturing Execution System = Empowerment" "	Jim Ricker
Evaporator Automation	Jose Danny" Milla, Marcelo Bellarde
Aseptic Bulk Storage - Critical Control Point Review	Jeffrey Raasch
Overview of Mechanical Harvesting in Florida Citrus	G.K. Brown

Volume XLVIII - 2002

The Impact of Brown Oil Recovery Systems on VOC and Wastewater Generation	Roger Waters
Pre-Extraction De-Oiling of Citrus Fruit by Roller Technology	Michael L. Suter
Flow and Heat Transfer Through Helix Tubular Coils	Dr. Gary C. Young, P.E.
Field Testing Improves Citrus Plant Biological Treatment	Jim Ashby, Tony Sobkowicz
Optimizing the Cost-Effectiveness of an Industrial Natural Gas Cogeneration Facility in a Regulated Electric Utility Environment	Mike Bedley
Continuing Education for Practicing Engineers	Jonathan F.K. Earle, Ph.D., P.E.

Volume XLIX - 2003

Brazil - Citrus Benchmarking	Alex Schuermans
Juice Room Environmental Control Through Misting	Cassio Luiz Ianni Zapaterra
Process Safety Management and Risk Management Program (PSM/RMP) Audits - Are You Prepared?	Martha Mead Ira
Tank Cleaning with Impact	Jan Hansen
Improved Agitators Applied to High Viscosity Fluids in Concentrated Juices	Edimilson Souza
Citrus Flavor Technologies	Elizabeth Berdis

Volume L - 2004

Packaging for Citrus Juices	Jean-Pierre Emond
Use of 3-A Standards in the Juice Industry	Timothy R. Rugh
World-Class Standards for Ammonia Refrigeration Systems	Joseph N. DeMarco
Integration of Statistical Process Control Data into Control and Information Systems for Production Optimization and Regulatory Compliance	Douglas C. Fair, Jim Redifer
Engineering Solutions for Lightning and Surge Protection	Paul Gates
Successful CIP Cleaning	Cliff Vaughn

Volume LI - 2005

Energy Savings and Opportunities in the Citrus Industry	Nathaniel Lewis Charles Thomas
New Systems in Plant Sanitation	Paul A. Johnson
New Processing Technology for Water as an Ingredient	Steven Moore
Bin Automation and Management	Rodney W. Swofford
Hurricane Panel Discussion – Lessons Learned	Barry L. Wilson, Moderator

Volume LII - 2006

Florida's Citrus Canker Eradication Program: Cost Benefit Analysis with an Example of an Endemic Citrus Canker Situation for Florida Processed Oranges	Ronald P. Muraro, Thomas H. Spreen, Marisa L. Zansler
Advanced System Controls and Energy Savings for Industrial Boilers	Daniel Willems
Citrus Oil Recovery – Using Modern "State-of-the-Art"Centrifuge Technology as Part of Economical Citrus Processing	Stefan Pecoroni
Machine Safety: Risk Assessment, Risk Reduction	Michael D. Ladd
Production Quality Improvements and Energy Cost	Mark Krisa
Reductions Associated with Compressed Air	David Voelker
Aseptic Ingredient Addition: Meeting the Demand for Better Tasting Orange Juice	Paul P. Brocker

Volume LIII - 2007

The Power of Insulation in the Citrus Industry	Ronald L. King
Manufacturing Intelligence: Understanding What Your Plant Is Trying To Tell You: Implementing A Web-Based Production Reporting System	Danny Milla
A Process and Economic Model for the Real-Time Optimization of Operating Profit in a Citrus Feedmill	Donald G. Gray, Jr.
Energy Efficient Lighting Design Criteria: Cost Effective Solutions in an Industrial Environment	Frank J. Agraz, Jr. John Maneri
The Development and Application of Resin Systems for the Treatment of Citrus Products Containing Pulp and Cloud	Tim Schofield Chris Miller
The Use and Misuse of 300 Series Austenitic Stainless Steels in the Citrus Industry	Richard O. Lewis
An Electrical Arc-Flash Hazard Analysis Primer: Reducing Arc-Flash Hard Exposures Through Engineering	John J. Kolak

Volume LIV - 2008

Commercial Ethanol Production from Cellulosic Biomass in South Florida	Craig Evans
Innovative Mechanical Water Treatment for Evaporative Condensers at a Citrus Facility	Phil Vella
Crossflow Filtration (Ceramic and Hollow Fiber Membranes) in the Citrus Industry	Rishi Sondhi
Non-Thermal Preservation of Fruit Juices	V.M. (Bala) Balasubramaniam

Energy and Environmental Conservation Programs Florida Utilities	Samuel Musura
Technical Training – Workforce Enrichment	Rob Clancey
Review of Dense Phase Carbon Dioxide Application to Citrus Juices	Murat Balaban

Volume LV – 2009

Update on HLB Progression and Current Research in Florida	Ron H. Brlansky
Converting Citrus Waste to Ethanol and Other Co-Products	Wilbur Widmer, Weiyang Zhou, Karel Grohmann
Sulfur Emission Reduction Project at Southern Gardens Citrus, Southern Gardens Citrus	Joe McDaniel Brendon Russ
Future Trends in Ammonia Refrigeration Design	Jeff Welch
Mechanical Harvesting's Impact on the Citrus Processing Industry	Reza Ehsani, Won Suk Lee, Elizabeth Northeimer
A Plant-Wide Insulation Assessment	Ronald L. King
Caustic Recovery Using Membrane Technology	Mike Grigus

Volume LVI – 2010

2010 Regulatory Update for Anydrous Ammonia Refrigeration Processes	Martha Mead Ira Ron H.
Independent Assessment of the Energy Savings, Environmental Improvements and Water Conservation of Emerging Non-Chemical Water Treatment Technologies	Michael Gravely, Bruce La Belle, John Balachandra
Chlorine Dioxide Sterilization of Bulk Aseptic Storage Tanks	Mark Morgan, Michael Burke
Rheological behavior of citrus pulp and its application to aseptic processing	Manuele Levati
Benefits and Techniques for Machine Condition Monitoring in Industrial Environments	Michael Denton

Volume LVII – 2012

Juice Debittering: Basic Science, Optimization, and Recent Advances	Fred Ghanem
Water Conservation and Wastewater Load Reduction in Citrus Processing	James Ashby
Press Liquor Filtration	Bob Johnston, PE
Flash Distillation Of D-Limonene In The Citrus Industry	Danny Milla
Applicability Of Greenhouse Gas (GHG) Regulations To Citrus Industry Operations	John M. Beath, PE
Orange Pulp Rheology	Jose Reyes De Corcuera
Ammonia Evaporative Condensers With Extended Surface Fins	Chris Dreher, PE
Defining And Implementing A Foundation For Energy Management	Paul Fowler

Volume LVIII - 2014

Engineered Air Curtains And Air Curtain Vestibules For High Traffic Cooler And Freezer Doorways	Pete Smith
Expanding The Horizon Of Citrus Fiber	Brock Lundberg
Tropicana's Optimization Of Landfill Methane Gas Use	Glenn Johnson

To Treat Or Not To Treat Citrus Wastewater: What Are The Alternatives?	Denise Johnston
Greening Panel Discussion - Our Future With Hlb: Where To From Here?	Dan King, Moderator